CIRCUMFERENCE

THINGS YOU SHOULD KNOW
(QUESTIONS AND ANSWERS)

By Rumi Michael Leigh

Introduction

I would like to thank you for purchasing this book, *"Circumference, things you should know (questions and answers)"*.

This book will help you understand, revise, and have a good general knowledge and understanding of the basics of circumference exercises.

I hope you enjoy it!

Table of Contents

Part 1: Circumference

Exercise 1

Questions

a) A circular table has a diameter of 120 cm. What is its original circumference?

b) A circular pond has an area of 154 sq. meters. What is its original circumference?

c) A circular pizza has an area of 314 sq.inches. What is its original circumference?

d) A circular track has a radius of 50 meters. What is its original circumference?

e) A circular mirror has a diameter of 60 cm. What is its original circumference?

Answers

a) A circular table has a diameter of 120 cm. What is its original circumference?

The diameter of the table is given as 120 cm

radius = 60 cm

$C = 2\pi r$

r is the radius of the circle

π is a mathematical constant approximately equal to 3.14

$C = 2 \times 3.14 \times 60$

C = 376.8 cm

Answer: the original circumference of the circular table is 376.8 cm

b) A circular pond has an area of 154 sq. meters. What is its original circumference?

$A = \pi r^2$

r is the radius of the circle

π is a mathematical constant approximately equal to 3.14

$154 = 3.14 \times r^2$

$r^2 = 154/3.14$

$r^2 \approx 49.0446 \ r \approx 7$

the radius of the pond is approximately 7 meters

C = 2 × 3.14 × 7

C = 43.96 meters

Answer: the original circumference of the circular pond is approximately 43.96 meters.

c) A circular pizza has an area of 314 sq.inches. What is its original circumference?

314 = 3.14 × r^2

r^2 = 314/3.14

r^2 ≈ 100

r ≈ 10

the radius of the pizza is approximately 10 inches

C = 2 × 3.14 × 10

C = 62.8 inches

Answer: the original circumference of the circular pizza is 62.8 inches.

d) A circular track has a radius of 50 meters. What is its original circumference?

C = 2 × 3.14 × 50

C = 314 meters

Answer: the original circumference of the circular track is 314 meters.

e) A circular mirror has a diameter of 60 cm. What is its original circumference?

radius = half of diameter = 30 cm

C = 2 × 3.14 × 30

C = 188.4 cm

Answer: the original circumference of the circular mirror is 188.4 cm

Exercise 2

Questions

a) A circular garden has an area of 314 sq. meters. What is its original circumference?

b) A circular plate has a diameter of 20 inches. What is its original circumference?

c) A circular pool has an area of 113.04 sq. meters. What is its original circumference?

d) A circular wheel has a radius of 30 cm. What is its original circumference?

e) A circular rug has a diameter of 8 feet. What is its original circumference?

Answers

a) A circular garden has an area of 314 sq. meters. What is its original circumference?

$314 = 3.14 \times r^2$

$r^2 = 314/3.14$

$r^2 \approx 100$

$r \approx 10$

$C = 2 \times 3.14 \times 10 \ C = 62.8$ meters

Answer: the original circumference of the circular garden is approximately 62.8 meters.

b) A circular plate has a diameter of 20 inches. What is its original circumference?

radius is half of diameter = 10 inches

$C = 2 \times 3.14 \times 10 \ C = 62.8$ inches

Answer: the original circumference of the circular plate is 62.8 inches.

c) A circular pool has an area of 113.04 sq. meters. What is its original circumference?

$113.04 = 3.14 \times r^2$

$r^2 = 113.04/3.14$

$r^2 \approx 36$

$r \approx 6$

$C = 2 \times 3.14 \times 6 \ C = 37.68$ meters

Answer: the original circumference of the circular pool is approximately 37.68 meters.

d) A circular wheel has a radius of 30 cm. What is its original circumference?

$C = 2 \times 3.14 \times 30 \ C = 188.4$ cm

Answer: the original circumference of the circular wheel is 188.4 cm.

e) A circular rug has a diameter of 8 feet. What is its original circumference?

Radius is half of diameter which is 4 feet.

$C = 2 \times 3.14 \times 4$

$C = 25.12$ feet

Answer: the original circumference of the circular rug is 25.12 feet.

Exercise 3

Questions

a) A circular table has a circumference of 62.8 inches. What is its original diameter?
b) A circular pizza has an original circumference of 75.4 cm. What is its original radius?
c) A circular trampoline has an area of 452.16 sq. feet. What is its original radius?
d) A circular pool has an original circumference of 150 meters. What is its original radius?
e) A circular hoop has an original circumference of 18.84 inches. What is its original radius?

Answers

a) A circular table has a circumference of 62.8 inches. What is its original diameter?
   $C = 2 \times 3.14 \times r$
   $62.8 = 2 \times 3.14 \times r$
   $r = 10$
   $d = 2 \times r$
   $d = 2 \times 10$
   $d = 20$ inches
   Answer: the original diameter of the circular table is 20 inches.
b) A circular pizza has an original circumference of 75.4 cm. What is its original radius?
   $C = 2 \times 3.14 \times r$
   $75.4 = 2 \times 3.14 \times r$
   $r \approx 12$
   Answer: the original radius of the circular pizza is approximately 12 cm.
c) A circular trampoline has an area of 452.16 sq. feet. What is its original radius?
   $452.16 = 3.14 \times r^2$

r^2 ≈ 144

r ≈ 12

Answer: the original radius of the trampoline is approximately 12 feet.

d) A circular pool has an original circumference of 150 meters. What is its original radius?

C = 2 × 3.14 × r

150 = 2 × 3.14 × r

r ≈ 23.9

Answer: the original radius of the circular pool is approximately 23.9 meters.

e) A circular hoop has an original circumference of 18.84 inches. What is its original radius?

C = 2 × 3.14 × r

18.84 = 2 × 3.14 × r

r ≈ 3

Answer: the original radius of the circular hoop is approximately 3 inches.

Exercise 4

Questions

a) A circular garden has an area of 78.5 square meters. What is its original circumference?

b) A circular track has an original circumference of 1.5 kilometers. What is its original radius?

c) A circular pie has an original diameter of 10 inches. What is its original circumference?

d) A circular swimming pool has an original circumference of 30 meters. What is its original area?

Answers

a) A circular garden has an area of 78.5 square meters. What is its original circumference?

A = 3.14 × r^2

78.5 = 3.14 × r^2

r ≈ 5

C = 2 × 3.14 × r

C = 2 × 3.14 × 5

C = 31.4 meters

Answer: the original circumference of the circular garden is 31.4 meters.

b) A circular track has an original circumference of 1.5 kilometers. What is its original radius?

C = 2 × 3.14 × r

1500 = 2 × 3.14 × r

r ≈ 238.7

Answer: the original radius of the circular track is approximately 238.7 meters.

c) A circular pie has an original diameter of 10 inches. What is its original circumference?

C = 3.14 × d

C = 3.14 × 10

C = 31.4 inches

Answer: the original circumference of the circular pie is 31.4 inches.

d) A circular swimming pool has an original circumference of 30 meters. What is its original area?

C = 2 × 3.14 × r

30 = 2 × 3.14 × r

r ≈ 4.77

A = 3.14 × r^2

A = 3.14 × (4.77)^2

A ≈ 71.35 square meters

Answer: the original area of the circular swimming pool is approximately 71.35 square meters.

Exercise 5

Questions

a) A circular pizza has an original area of 100 square inches. What is its original diameter?

b) A circular table has an original area of 36 square feet. What is its original radius?

c) A circular well has an original diameter of 6 meters. What is its original circumference?

d) A circular tire has an original radius of 14 inches. What is its original circumference?

e) A circular pond has an original area of 50 square meters. What is its original circumference?

Answers

a) A circular pizza has an original area of 100 square inches. What is its original diameter?

A = 3.14 × r^2

100 = 3.14 × r^2

r ≈ 5.64

d = 2 × r

d = 2 × 5.64

d ≈ 11.28 inches

Answer: the original diameter of the circular pizza is approximately 11.28 inches.

b) A circular table has an original area of 36 square feet. What is its original radius?

A = 3.14 × r^2

36 = 3.14 × r^2

r ≈ 3

Answer: the original radius of the circular table is approximately 3 feet.

c) A circular well has an original diameter of 6 meters. What is its original circumference?

C = 3.14 × d

C = 3.14 × 6

C = 18.84 meters

Answer: the original circumference of the circular well is 18.84 meters.

d) A circular tire has an original radius of 14 inches. What is its original circumference?

$C = 2 \times 3.14 \times r$

$C = 2 \times 3.14 \times 14$

$C \approx 87.96$ inches

Answer: the original circumference of the circular tire is approximately 87.96 inches.

e) A circular pond has an original area of 50 square meters. What is its original circumference?

$A = 3.14 \times r^2$

$50 = 3.14 \times r^2$

$r \approx 3.99$

$C = 2 \times 3.14 \times r$

$C = 2 \times 3.14 \times 3.99$

$C \approx 25$

Part 2: Circumference

Exercise 1

Questions

a) A circular field has an original circumference of 100 meters. What is its original radius?

b) A circular rug has an original diameter of 8 feet. What is its original area?

c) A circular window has an original circumference of 36 inches. What is its original diameter?

d) A circular garden has an original area of 25 square meters. What is its original diameter?

e) A circular plate has an original radius of 10 centimeters. What is its original area?

Answers

a) A circular field has an original circumference of 100 meters. What is its original radius?

$C = 2 \times 3.14 \times r$

$100 = 2 \times 3.14 \times r$

$r \approx 15.92$

Answer: the original radius of the circular field is approximately 15.92 meters.

b) A circular rug has an original diameter of 8 feet. What is its original area?

$r = d/2$

$r = 8/2$

$r = 4$

$A = 3.14 \times r^2$

$A = 3.14 \times 4^2$

$A \approx 50.24$ square feet

Answer: the original area of the circular rug is approximately 50.24 square feet.

c) A circular window has an original circumference of 36 inches. What is its original diameter?

C = 3.14 × d

36 = 3.14 × d

d ≈ 11.46

Answer: the original diameter of the circular window is approximately 11.46 inches.

d) A circular garden has an original area of 25 square meters. What is its original diameter?

A = 3.14 × r^2

25 = 3.14 × r^2

r ≈ 2.82

d = 2 × r

d = 2 × 2.82

d ≈ 5.64 meters

Answer: the original diameter of the circular garden is approximately 5.64 meters.

e) A circular plate has an original radius of 10 centimeters. What is its original area?

A = 3.14 × r^2

A = 3.14 × 10^2

A ≈ 314 square centimeters

Answer: the original area of the circular plate is approximately 314 square centimeters.

Exercise 2

Questions

a) A circular track has an original circumference of 400 meters. What is its original radius?

b) A circular pond has an original diameter of 8 meters. What is its original area?

r =4 (half of the diameter)

c) A circular pizza has an original circumference of 24 inches. What is its original diameter?

d) A circular table has an original radius of 3 feet. What is its original area?

e) A circular plate has an original circumference of 12 inches. What is its original radius?

Answers

a) A circular track has an original circumference of 400 meters. What is its original radius?

C = 2 × 3.14 × r

400 = 2 × 3.14 × r

r ≈ 63.66

Answer: the original radius of the circular track is approximately 63.66 meters.

b) A circular pond has an original diameter of 8 meters. What is its original area?

r =4 (half of the diameter)

A = 3.14 × r^2

A = 3.14 × 4^2

A ≈ 50.24 square meters

Answer: the original area of the circular pond is approximately 50.24 square meters.

c) A circular pizza has an original circumference of 24 inches. What is its original diameter?

C = 3.14 × d

24 = 3.14 × d

d ≈ 7.64

Answer: the original diameter of the circular pizza is approximately 7.64 inches.

d) A circular table has an original radius of 3 feet. What is its original area?

A = 3.14 × r^2

A = 3.14 × 3^2

A ≈ 28.26 square feet

Answer: the original area of the circular table is approximately 28.26 square feet.

e) A circular plate has an original circumference of 12 inches. What is its original radius?

C = 2 × 3.14 × r

12 = 2 × 3.14 × r

r ≈ 1.91

Answer: the original radius of the circular plate is approximately 1.91 inches.

Exercise 3

Questions

a) A circular rug has an original area of 100 square feet. What is its original radius?
b) A circular pool has an original diameter of 10 meters. What is its original area?
c) A circular clock has an original circumference of 36 centimeters. What is its original diameter?
d) A circular garden has an original radius of 6 meters. What is its original area?
e) A circular plate has an original diameter of 8 inches. What is its original area?

Answers

a) A circular rug has an original area of 100 square feet. What is its original radius?

$A = 3.14 \times r^2$

$100 = 3.14 \times r^2$

$r ≈ 5.64$

Answer: the original radius of the circular rug is approximately 5.64 feet.

b) A circular pool has an original diameter of 10 meters. What is its original area?

$r = d/2$

$r = 10/2$

$r = 5$

$A = 3.14 \times r^2$

$A = 3.14 \times 5^2$

$A ≈ 78.5$ square meters

Answer: the original area of the circular pool is approximately 78.5 square meters.

c) A circular clock has an original circumference of 36 centimeters. What is its original diameter?

$C = 3.14 \times d$

$36 = 3.14 \times d$

d ≈ 11.46

Answer: the original diameter of the circular clock is approximately 11.46 centimeters.

d) A circular garden has an original radius of 6 meters. What is its original area?

A = 3.14 × r^2

A = 3.14 × 6^2

A ≈ 113.04 square meters

Answer: the original area of the circular garden is approximately 113.04 square meters.

e) A circular plate has an original diameter of 8 inches. What is its original area?

r = d/2

r = 8/2

r = 4

A = 3.14 × r^2

A = 3.14 × 4^2

A ≈ 50.24 square inches

Answer: the original area of the circular plate is approximately 50.24 square inches.

Exercise 4

Questions

a) A circular pool has an original area of 314 square feet. What is its original radius?

b) A circular clock has an original diameter of 12 centimeters. What is its original circumference?

c) A circular garden has an original circumference of 24 meters. What is its original radius?

d) A circular plate has an original area of 78.5 square inches. What is its original diameter?

e) A circular pool has an original radius of 8 meters. What is its original area?

Answers

a) A circular pool has an original area of 314 square feet. What is its original radius?

A = 3.14 × r^2

314 = 3.14 × r^2

r ≈ 10

Answer: the original radius of the circular pool is approximately 10 feet.

b) A circular clock has an original diameter of 12 centimeters. What is its original circumference?

r = d/2

r = 12/2

r = 6

C = 2 × 3.14 × r

C = 2 × 3.14 × 6

C ≈ 37.68 centimeters

Answer: the original circumference of the circular clock is approximately 37.68 centimeters.

c) A circular garden has an original circumference of 24 meters. What is its original radius?

C = 2 × 3.14 × r

24 = 2 × 3.14 × r

r ≈ 3.82

Answer: the original radius of the circular garden is approximately 3.82 meters.

d) A circular plate has an original area of 78.5 square inches. What is its original diameter?

A = 3.14 × r^2

78.5 = 3.14 × r^2

r ≈ 5

d = 2 × r

d = 2 × 5

d = 10

Answer: the original diameter of the circular plate is 10 inches.

e) A circular pool has an original radius of 8 meters. What is its original area?

A = 3.14 × r^2

A = 3.14 × 8^2

A ≈ 200.96 square meters

Answer: the original area of the circular pool is approximately 200.96 square meters

Exercise 5

Questions

a) A circular clock has an original circumference of 18 inches. What is its original radius?

b) A circular garden has an original area of 50 square meters. What is its original circumference?

c) A circular pool has an original area of 113.1 square meters. What is its original diameter?

Answers

a) A circular clock has an original circumference of 18 inches. What is its original radius?

C = 2 × 3.14 × r

18 = 2 × 3.14 × r

r ≈ 2.87

Answer: the original radius of the circular clock is approximately 2.87 inches.

b) A circular garden has an original area of 50 square meters. What is its original circumference?

A = 3.14 × r^2

50 = 3.14 × r^2

r ≈ 3.18

C = 2 × 3.14 × r

C = 2 × 3.14 × 3.18

C ≈ 20.01 meters

Answer: the original circumference of the circular garden is approximately 20.01 meters.

c) A circular pool has an original area of 113.1 square meters. What is its original diameter?

$A = 3.14 \times r^2$

$113.1 = 3.14 \times r^2$

$r \approx 6$

$d = 2 \times r$

$d = 2 \times 6$

$d = 12$

Answer: the original diameter of the circular pool is 12 meters.

Part 3: Circumference

Exercise 1

Questions

a) A circular clock has an original radius of 5 inches. What is its original area?
b) A circular garden has an original circumference of 28.26 meters. What is its original radius?
c) A circular plate has an original area of 201.06 square inches. What is its original radius?
d) A circular pool has an original circumference of 50.24 meters. What is its original radius?
e) A circular clock has an original diameter of 12 inches. What is its original circumference?

Answers

a) A circular clock has an original radius of 5 inches. What is its original area?
   A = 3.14 × r^2
   A = 3.14 × 5^2
   A ≈ 78.5 square inches
   Answer: the original area of the circular clock is approximately 78.5 square inches.

b) A circular garden has an original circumference of 28.26 meters. What is its original radius?
   C = 2 × 3.14 × r 28.26 = 2 × 3.14 × r r ≈ 4.5
   Answer: the original radius of the circular garden is approximately 4.5 meters.

c) A circular plate has an original area of 201.06 square inches. What is its original radius?
   A = 3.14 × r^2
   201.06 = 3.14 × r^2
   r ≈ 8
   Answer: the original radius of the circular plate is approximately 8 inches.

d) A circular pool has an original circumference of 50.24 meters. What is its original radius?

C = 2 × 3.14 × r

50.24 = 2 × 3.14 × r

r ≈ 8

Answer: the original radius of the circular pool is approximately 8 meters.

e) A circular clock has an original diameter of 12 inches. What is its original circumference?

C = 3.14 × d

C = 3.14 × 12

C ≈ 37.68 inches

Answer: the original circumference of the circular clock is approximately 37.68 inches.

Exercise 2

Questions

a) A circular garden has an original diameter of 20 meters. What is its original circumference?

b) A circular plate has an original circumference of 62.8 inches. What is its original radius?

c) A circular pool has an original diameter of 15 meters. What is its original area?

d) A circular clock has an original area of 28.26 square inches. What is its original radius?

Answers

a) A circular garden has an original diameter of 20 meters. What is its original circumference?

C = 3.14 × d

C = 3.14 × 20

C ≈ 62.8 meters

Answer: the original circumference of the circular garden is approximately 62.8 meters.

b) A circular plate has an original circumference of 62.8 inches. What is its original radius?

C = 2 × 3.14 × r

62.8 = 2 × 3.14 × r

r ≈ 10

Answer: the original radius of the circular plate is approximately 10 inches.

c) A circular pool has an original diameter of 15 meters. What is its original area?

A = 3.14 × (d/2)^2

A = 3.14 × (15/2)^2

A ≈ 176.62 square meters

Answer: the original area of the circular pool is approximately 176.62 square meters.

d) A circular clock has an original area of 28.26 square inches. What is its original radius?

A = 3.14 × r^2

28.26 = 3.14 × r^2

r ≈ 3

Answer: the original radius of the circular clock is approximately 3 inches.

Exercise 3

Questions

a) A circular plate has an original diameter of 18 inches. What is its original area?
b) A circular pool has an original area of 113.04 square meters. What is its original radius?

Answers

a) A circular plate has an original diameter of 18 inches. What is its original area?

A = 3.14 × (d/2)^2

A = 3.14 × (18/2)^2

A ≈ 254.34 square inches

Answer: the original area of the circular plate is approximately 254.34 square inches.

b) A circular pool has an original area of 113.04 square meters. What is its original radius?

A = 3.14 × r^2

113.04 = 3.14 × r^2

r ≈ 6

Answer: the original radius of the circular pool is approximately 6 meters.

Exercise 4

Questions

a) A circular pool has an original radius of 12 meters. What is its original area?

b) A circular clock has an original radius of 5 inches. What is its original area?

c) A circular garden has an original radius of 8 meters

d) A circular plate has an original circumference of 62.8 inches. What is its original area?

e) A circular pool has an original area of 50.24 square meters. What is its original circumference?

Answers

a) A circular pool has an original radius of 12 meters. What is its original area?

A = 3.14 × r^2

A = 3.14 × 12^2

A ≈ 452.16 square meters

Answer: the original area of the circular pool is approximately 452.16 square meters.

b) A circular clock has an original radius of 5 inches. What is its original area?

A = 3.14 × r^2

A = 3.14 × 5^2

A ≈ 78.5 square inches

Answer: the original area of the circular clock is approximately 78.5 square inches.

c) A circular garden has an original radius of 8 meters

A = 3.14 × r^2

A = 3.14 × 8^2

A ≈ 200.96 square meters

Answer: the original area of the circular garden is approximately 200.96 square meters.

d) A circular plate has an original circumference of 62.8 inches. What is its original area?

C = 2 × 3.14 × r

62.8 = 2 × 3.14 × r

r ≈ 10

A = 3.14 × r^2

A = 3.14 × 10^2

A ≈ 314 square inches

Answer: the original area of the circular plate is approximately 314 square inches.

e) A circular pool has an original area of 50.24 square meters. What is its original circumference?

A = 3.14 × r^2

50.24 = 3.14 × r^2

r ≈ 4

C = 2 × 3.14 × r

C = 2 × 3.14 × 4

C ≈ 25.12 meters

Answer: the original circumference of the circular pool is approximately 25.12 meters.

Exercise 5

Questions

a) A circular garden has an original area of 154 square feet. If the garden is expanded so that its new radius is twice its original radius, what is the new area of the garden?

b) A circular garden has an original circumference of 62.8 meters. If the garden is expanded so that its new circumference is 100 meters, what is the new radius of the garden?

c) A circular pool has an original circumference of 37.68 feet. What is its original area?

d) A circular plate has an original circumference of 25.12 inches. If the plate is expanded so that its new radius is 50% larger than its original radius, what is the new circumference of the plate?

e) A circular garden has an original circumference of 28 meters. If the garden is reduced so that its new radius is half its original radius, what is the new circumference of the garden?

Answers

a) A circular garden has an original area of 154 square feet. If the garden is expanded so that its new radius is twice its original radius, what is the new area of the garden?

$A = 3.14 \times r^2$ $154 = 3.14 \times r^2$ $r \approx 7$

The new radius = $2 \times 7 = 14$ feet.

$A = 3.14 \times r^2$

$A = 3.14 \times 14^2$

$A \approx 615.44$ square feet

Answer: the new area of the circular garden is approximately 615.44 square feet.

b) A circular garden has an original circumference of 62.8 meters. If the garden is expanded so that its new circumference is 100 meters, what is the new radius of the garden?

$C = 2 \times 3.14 \times r$

$62.8 = 2 \times 3.14 \times r$

$r \approx 10$

$C = 2 \times 3.14 \times r$

$100 = 2 \times 3.14 \times r$

$r \approx 15.92$

Answer: the new radius of the circular garden is approximately 15.92 meters.

c) A circular pool has an original circumference of 37.68 feet. What is its original area?

$C = 2 \times 3.14 \times r$

$37.68 = 2 \times 3.14 \times r$

$r \approx 6$

$A = 3.14 \times r^2$

$A = 3.14 \times 6^2$

$A \approx 113.04$ square feet

Answer: the original area of the circular pool is approximately 113.04 square feet.

d) A circular plate has an original circumference of 25.12 inches. If the plate is expanded so that its new radius is 50% larger than its original radius, what is the new circumference of the plate?

$C = 2 \times 3.14 \times r$

$25.12 = 2 \times 3.14 \times r$

$r \approx 4$

The new radius of the plate is 50% larger than the original radius, which means it is 1.5 times the original radius. So, the new radius is $1.5 \times 4 = 6$ inches.

$C = 2 \times 3.14 \times r$

$C = 2 \times 3.14 \times 6$

$C \approx 37.68$ inches

Answer: the new circumference of the circular plate is approximately 37.68 inches.

e) A circular garden has an original circumference of 28 meters. If the garden is reduced so that its new radius is half its original radius, what is the new circumference of the garden?

$C = 2 \times 3.14 \times r$

$28 = 2 \times 3.14 \times r$

$r \approx 4.46$

The new radius of the garden is half the original radius, which means it is $0.5 \times 4.46 = 2.23$ meters.

C = 2 × 3.14 × r

C = 2 × 3.14 × 2.23

C ≈ 14 meters

Therefore, the new circumference of the circular garden is approximately 14 meters.

Part 4: Circumference

Exercise 1

Questions

a) A circular plate has an original circumference of 22 inches. If the plate is reduced so that its new area is half its original area, what is the new radius of the plate?

b) A circular garden has an original circumference of 48 feet. If the garden is reduced so that its new radius is 75% of its original radius, what is the new circumference of the garden?

c) A circular swimming pool has an original circumference of 60 meters. If the depth of the pool is doubled and the pool is expanded so that its new area is twice its original area, what is the new circumference of the pool?

Answers

a) A circular plate has an original circumference of 22 inches. If the plate is reduced so that its new area is half its original area, what is the new radius of the plate?

$C = 2 \times 3.14 \times r$

$22 = 2 \times 3.14 \times r$

$r \approx 3.5$

$A = 3.14 \times r^2$

$A = 3.14 \times 3.5^2$

$A \approx 38.47$ square inches

The new area of the plate is half the original area, which means it is $0.5 \times 38.47 = 19.24$ square inches.

$A = 3.14 \times r^2$

$19.24 = 3.14 \times r^2$

$r \approx 2.21$

Answer: the new radius of the circular plate is approximately 2.21 inches.

b) A circular garden has an original circumference of 48 feet. If the garden is reduced so that its new radius is 75% of its original radius, what is the new circumference of the garden?

C = 2 × 3.14 × r

48 = 2 × 3.14 × r

r ≈ 7.64

The new radius of the garden is 75% of the original radius, which means it is 0.75 × 7.64 = 5.73 feet.

C = 2 × 3.14 × r

C = 2 × 3.14 × 5.73

C ≈ 36 feet

Answer: the new circumference of the circular garden is approximately 36 feet.

c) A circular swimming pool has an original circumference of 60 meters. If the depth of the pool is doubled and the pool is expanded so that its new area is twice its original area, what is the new circumference of the pool?

C = 2 × 3.14 × r

60 = 2 × 3.14 × r

r ≈ 9.55

The original area of the pool can be found using the formula:

A = 3.14 × r^2

A = 3.14 × 9.55^2

A ≈ 286.77 square meters

If the depth of the pool is doubled, then the new area of the pool will be:

2A = 2 × 286.77

2A = 573.54 square meters

A = 3.14 × r^2

573.54 = 3.14 × r^2

r ≈ 13.57

C = 2 × 3.14 × r

C = 2 × 3.14 × 13.57

C ≈ 85.25 meters

Answer: the new circumference of the circular swimming pool is approximately 85.25 meters.

Exercise 2

Questions

a) A circular plate has an original area of 49 square inches. If the plate is expanded so that its new radius is twice its original radius, what is the new area of the plate?

b) A circular pizza has an original diameter of 16 inches. If the pizza is cut into 8 equal slices, what is the perimeter of each slice?

c) A circular water tank has an original diameter of 10 meters. If the water tank is filled to a depth of 8 meters, what is the volume of water in the tank?

d) A circular track has an original circumference of 500 meters. If the track is widened so that its new radius is 1.5 times its original radius, what is the new area of the track?

Answers

a) A circular plate has an original area of 49 square inches. If the plate is expanded so that its new radius is twice its original radius, what is the new area of the plate?

$A = 3.14 \times r^2$

$49 = 3.14 \times r^2$

$r \approx 3.96$

The new radius of the plate is twice the original radius, which means it is $2 \times 3.96 = 7.92$ inches.

$A = 3.14 \times r^2$

$A = 3.14 \times 7.92^2$

$A \approx 197.93$ square inches

Answer: the new area of the circular plate is approximately 197.93 square inches.

b) A circular pizza has an original diameter of 16 inches. If the pizza is cut into 8 equal slices, what is the perimeter of each slice?

$C = 2 \times 3.14 \times r$

$C = 2 \times 3.14 \times 8$

C ≈ 50.24 inches

When the pizza is cut into 8 equal slices, each slice will have an angle of 45 degrees, since 360 degrees divided by 8 equals 45 degrees. The perimeter of each slice can be found by multiplying the arc length of the slice by 2, since the slice has a curved edge on both sides.

The arc length of each slice can be found by multiplying the circumference of the pizza by the angle of the slice in degrees divided by 360 degrees:

Arc length = C × (angle/360)

Arc length = 50.24 × (45/360)

Arc length ≈ 6.28 inches

Answer: the perimeter of each slice of pizza is approximately 2 × 6.28 = 12.56 inches.

c) A circular water tank has an original diameter of 10 meters. If the water tank is filled to a depth of 8 meters, what is the volume of water in the tank?

V = π × r^2 × h

V = 3.14 × 5^2 × 8

V ≈ 628.32 cubic meters

Answer: the volume of water in the circular water tank is approximately 628.32 cubic meters.

d) A circular track has an original circumference of 500 meters. If the track is widened so that its new radius is 1.5 times its original radius, what is the new area of the track?

C = 2 × 3.14 × r

500 = 2 × 3.14 × r

r ≈ 79.62

The new radius of the track is 1.5 times the original radius, which means it is 1.5 × 79.62 = 119.43 meters.

A = 3.14 × r^2

A = 3.14 × 119.43^2

A ≈ 44709.81 square meters

Answer: the new area of the circular track is approximately 44709.81 square meters.

Exercise 3

Questions

a) A circular garden has an original diameter of 12 meters. If the garden is enlarged so that its new radius is twice its original radius, what is the new circumference of the garden?

b) A circular track has an original circumference of 200 meters. If the track is reduced so that its new radius is half its original radius, what is the new area of the track?

c) A circular swimming pool has an original radius of 8 meters. If the pool is filled to a depth of 4 meters, what is the volume of water in the pool?

d) A circular plate has an original radius of 10 inches. If the plate is reduced so that its new diameter is 12 inches, what is the new area of the plate?

Answers

a) A circular garden has an original diameter of 12 meters. If the garden is enlarged so that its new radius is twice its original radius, what is the new circumference of the garden?

The radius is half of the diameter, or 6 meters.

The new radius of the garden is twice the original radius, which means it is $2 \times 6 = 12$ meters.

$C = 2 \times 3.14 \times r$

$C = 2 \times 3.14 \times 12$

$C \approx 75.36$ meters

Answer: the new circumference of the circular garden is approximately 75.36 meters.

b) A circular track has an original circumference of 200 meters. If the track is reduced so that its new radius is half its original radius, what is the new area of the track?

$C = 2 \times 3.14 \times r$

$200 = 2 \times 3.14 \times r$

$r \approx 31.85$ meters

The new radius of the track is half the original radius, which means it is 0.5 × 31.85 = 15.92 meters.

A = 3.14 × r^2

A = 3.14 × 15.92^2

A ≈ 1005.03 square meters

Answer: the new area of the circular track is approximately 1005.03 square meters.

c) A circular swimming pool has an original radius of 8 meters. If the pool is filled to a depth of 4 meters, what is the volume of water in the pool?

V = π × r^2 × h

V = 3.14 × 8^2 × 4

V ≈ 803.84 cubic meters

Answer: the volume of water in the circular swimming pool is approximately 803.84 cubic meters.

d) A circular plate has an original radius of 10 inches. If the plate is reduced so that its new diameter is 12 inches, what is the new area of the plate?

A = 3.14 × r^2

A = 3.14 × 10^2

A ≈ 314.00 square inches

The new diameter of the plate is given as 12 inches, which means the new radius is half of the diameter, or 6 inches.

A = 3.14 × r^2

A = 3.14 × 6^2

A ≈ 113.04 square inches

Answer: the new area of the circular plate is approximately 113.04 square inches.

Exercise 4

Questions

a) A circular flower bed has an original circumference of 24 meters. If the flower bed is enlarged so that its new diameter is 16 meters, what is the new area of the flower bed?

b) A circular pond has an original radius of 15 feet. If the pond is filled to a depth of 5 feet, what is the volume of water in the pond?

c) A circular table has an original diameter of 1 meter. If the table is reduced so that its new radius is half its original radius, what is the new circumference of the table?

d) A circular pizza has an original radius of 12 inches. If the pizza is cut into 6 equal slices, what is the area of each slice?

e) A circular plate has an original circumference of 36 centimeters. If the plate is reduced so that its new radius is half its original radius, what is the new area of the plate?

Answers

a) A circular flower bed has an original circumference of 24 meters. If the flower bed is enlarged so that its new diameter is 16 meters, what is the new area of the flower bed?

$C = 2 \times 3.14 \times r$

$24 = 2 \times 3.14 \times r$

$r \approx 3.82$ meters

The new diameter of the flower bed is given as 16 meters, which means the new radius is half of the diameter, or 8 meters.

$A = 3.14 \times r^2$

$A = 3.14 \times 8^2$

$A \approx 200.96$ square meters

Answer: the new area of the circular flower bed is approximately 200.96 square meters.

b) A circular pond has an original radius of 15 feet. If the pond is filled to a depth of 5 feet, what is the volume of water in the pond?

$V = \pi \times r^2 \times h$

$V = 3.14 \times 15^2 \times 5$

$V \approx 3533.00$ cubic feet

Answer: the volume of water in the circular pond is approximately 3533.00 cubic feet.

c) A circular table has an original diameter of 1 meter. If the table is reduced so that its new radius is half its original radius, what is the new circumference of the table?

The original diameter of the table is given as 1 meter, which means the original radius is half of the diameter, or 0.5 meters.

The new radius of the table is half the original radius, which means it is 0.5 × 0.5 = 0.25 meters.

$C = 2 \times 3.14 \times r$

$C = 2 \times 3.14 \times 0.25$

$C \approx 1.57$ meters

Answer: the new circumference of the circular table is approximately 1.57 meters.

d) A circular pizza has an original radius of 12 inches. If the pizza is cut into 6 equal slices, what is the area of each slice?

$A = 3.14 \times r^2$

$A = 3.14 \times 12^2$

$A \approx 452.16$ square inches

If the pizza is cut into 6 equal slices, each slice will have an angle of 360/6 = 60 degrees.

$A = (\theta/360) \times 3.14 \times r^2$

$A = (60/360) \times 3.14 \times 12^2$

$A \approx 75.40$ square inches

Answer: each slice of the circular pizza has an area of approximately 75.40 square inches.

e) A circular plate has an original circumference of 36 centimeters. If the plate is reduced so that its new radius is half its original radius, what is the new area of the plate?

$C = 2 \times 3.14 \times r$

$36 = 2 \times 3.14 \times r$

$r \approx 5.73$ centimeters

The new radius of the plate is half the original radius, which means it is 5.73/2 = 2.86 centimeters.

$A = 3.14 \times r^2$

$A = 3.14 \times 2.86^2$

A ≈ 25.68 square centimeters

Answer: the new area of the circular plate is approximately 25.68 square centimeters.

Part 5: Circumference

Exercise 1

Questions

a) A wheel has a circumference of 120 centimeters. What is the diameter of the wheel?

b) The circumference of a circle is 18π centimeters. What is the radius of the circle?

c) A circular swimming pool has a circumference of 66 feet. What is the diameter of the pool?

d) The circumference of a wheel is 60 meters. How many revolutions will the wheel make if it travels a distance of 480 meters?

e) The diameter of a circular rug is 8 feet. What is the circumference of the rug?

Answers

a) A wheel has a circumference of 120 centimeters. What is the diameter of the wheel?

$C = \pi d$

$120 = \pi d$

$d = 120/\pi$

We can simplify this expression by approximating π to 3.14:

$d \approx 120/3.14$

$d \approx 38.22$

Answer: the diameter of the wheel is approximately 38.22 centimeters.

b) The circumference of a circle is 18π centimeters. What is the radius of the circle?

$C = 2\pi r$

$18\pi = 2\pi r$

$r = 9$

Answer: the radius of the circle is 9 centimeters.

c) A circular swimming pool has a circumference of 66 feet. What is the diameter of the pool?

$C = \pi d$

$66 = \pi d$

$d = 66/\pi$

We can simplify this expression by approximating π to 3.14

$d \approx 66/3.14$

$d \approx 21.02$

Answer: the diameter of the pool is approximately 21.02 feet.

d) The circumference of a wheel is 60 meters. How many revolutions will the wheel make if it travels a distance of 480 meters?

Each revolution of the wheel covers a distance equal to its circumference.

Number of revolutions = Distance traveled / Circumference of wheel Number of revolutions = 480 / 60 Number of revolutions = 8

Answer: the wheel will make 8 revolutions if it travels a distance of 480 meters.

e) The diameter of a circular rug is 8 feet. What is the circumference of the rug?

$C = \pi d$

$C = \pi(8)$

$C = 8\pi$

Answer: the circumference of the rug is 8π feet.

Exercise 2

Questions

a) A bicycle tire has a circumference of 84 inches. What is the diameter of the tire?
b) The circumference of a circular table is 36 feet. What is the diameter of the table?
c) A circular pizza has a diameter of 16 inches. What is the circumference of the pizza?
d) The circumference of a circular garden is 50 meters. What is the radius of the garden?

Answers

a) A bicycle tire has a circumference of 84 inches. What is the diameter of the tire?

$C = \pi d$

$84 = \pi d$

$d = 84/\pi$

We can simplify this expression by approximating π to 3.14:

$d \approx 84/3.14$

$d \approx 26.75$

Answer: the diameter of the tire is approximately 26.75 inches.

b) The circumference of a circular table is 36 feet. What is the diameter of the table?

$C = \pi d$

$36 = \pi d$

$d = 36/\pi$

We can simplify this expression by approximating π to 3.14:

$d \approx 36/3.14$

$d \approx 11.46$

Answer: the diameter of the table is approximately 11.46 feet.

c) A circular pizza has a diameter of 16 inches. What is the circumference of the pizza?

$C = \pi d$

$C = \pi(16)$

$C = 16\pi$

Answer: the circumference of the pizza is 16π inches.

d) The circumference of a circular garden is 50 meters. What is the radius of the garden?

$C = 2\pi r$

$50 = 2\pi r$

$r = 25/\pi$

We can simplify this expression by approximating π to 3.14:

$r \approx 25/3.14$

$r \approx 7.96$

Answer: the radius of the garden is approximately 7.96 meters.

Conclusion

Thank you once again for purchasing this book. I hope it has helped you in your journey to understand the basics of circumference.

Please, if you learnt something from this book, I would like you to leave a review. It'd be appreciated.

Thank you.